Bibliographic information published by the German National Library:

The German National Library lists this publication in the National Bibliography; detailed bibliographic data are available on the Internet at http://dnb.dnb.de .

Imprint:

Print and binding: Books on Demand GmbH, Norderstedt Germany
ISBN: 9783668455412

This book at GRIN:

http://www.grin.com/en/e-book/366501/systems-and-processes-defined-by-the-substances-matter-energy-and-information

Alexander Mircescu

Systems and Processes Defined by the Substances Matter, Energy, and Information in the Existing Forms Space, Time, and Causality

GRIN Publishing

Systems and Processes Defined by the Substances Matter, Energy, and Information in the Existing Forms Space, Time, and Causality

Dr. Alexander Mircescu, Munich, Germany

I. Definition of Systems and Processes

Automation systems are the most general systems known in engineering, since they couple the management of matter, energy, and information in space, time, and causality. Indeed, such systems define entire production processes where materials are processed, transported, and stored. The production processes of materials require (mostly electrical) energy needed for the operating machines, such that energy has also to be transformed, transported, and stored. The machines are controlled by computers, such that information flows are also present, implying that information has also to be processed, communicated among the operating machines, and stored. In order to formalize the description of such automation processes we will define a system and a process in a deductive manner in this chapter. This definition will appear astonishing at this step, but will be clarified in the following chapters explaining what information and causality are, and how they behave in physics together with matter, energy, space, and time [6].

Definition 1 [6]: ***System***: A system is a ten dimensional vector consisting of 3 dimensions of translational and rotational space, $\left((x,\varphi_{yz}),(y,\varphi_{zx}),(z,\varphi_{xy})\right)$, 3 complementary dimensions of space given by the overall momenta $\left((p_x,L_{yz}),(p_y,L_{zx}),(p_z,L_{xy})\right)$, 1 dimension of time (t), 1 complementary dimension of time given by the energy (E), 1 dimension of causality (k), and 1 complementary dimension of causality given by the information (I):

$$System = \begin{pmatrix} (x,\varphi_{yz}) \\ (y,\varphi_{zx}) \\ (z,\varphi_{xy}) \\ (p_x,L_{yz}) \\ (p_y,L_{zx}) \\ (p_z,L_{xy}) \\ t \\ E \\ k \\ I \end{pmatrix}$$

Definition 2 [6]: ***Process***: A process is a nine dimensional entity consisting of a 3×3 matrix:

$$Process = \begin{pmatrix} Matter_{Processing} & Matter_{Transport} & Matter_{Storage} \\ Energy_{Processing} & Energy_{Transport} & Energy_{Storage} \\ Information_{Processing} & Information_{Transport} & Information_{Storage} \end{pmatrix}$$

such that each element of the matrix defines one dimension.

II. Complementary Coordinates

0.1 Momentum and Space

In physics there are two possibilities to show the complementarity between momentum $\vec{p}$ and space $\vec{r}$: The differential operator and the Fourier methodology.

0.1.1 Differential Operator Methodology

In quantum mechanics momentum $\vec{p}$ is a differential operator, defined as [1]:

$$p_x = -i\hbar \frac{\partial}{\partial x}$$

with p_x as the component of $\vec{p}$ in the x direction of space; i as the imaginary unit ($i^2 = -1$) of the complex numbers ($\mathbb{C}$); $\hbar$ as the reduced Planck constant; and $\frac{\partial}{\partial x}$ as the partial derivative in the x direction of space. The momentum operator maps the wave function to another function. If this new function is a constant multiplied by the original wave function, then the momentum operator defines the eigenvalue, and the wave function defines the eigenfunction [1]. The set of eigenvalues of the momentum operator are possible results measured in an experiment [1].

In quantum mechanics the spatial position $\vec{r}$ is an operator that corresponds to the spatial position observable of a particle [1]. In momentum space, the position operator is defined as [1]:

$$x = i\hbar \frac{\partial}{\partial p_x}$$

with x as the x direction of space; i as the imaginary unit ($i^2 = -1$) of the complex numbers ($\mathbb{C}$); $\hbar$ as the reduced Planck constant; and $\frac{\partial}{\partial p_x}$ as the partial derivative in the x direction of momentum.

By using the differential operators for momentum and spatial position one obtains the canonical commutation relation [1]:

$$[x, p_x] = xp_x - p_x x = i\hbar$$

This leads to the uncertainty relation between momentum and space as postulated by Werner Heisenberg:

$$\Delta p_x \Delta x \approx h$$

Corollary 0.1: Hence, momentum and space can be regarded as complementary by the differential operator methodology, in the sense that they are inverse differential operators.

0.1.2 Fourier Methodology

According to [7] we can approximate a function $f(x)$ by:

$$f(x) = \int_{-\infty}^{\infty} F(\xi) e^{i2\pi\xi x} \, d\xi$$

with

$$F(\xi) = \int_{-\infty}^{\infty} f(x) e^{-i2\pi\xi x} \, dx$$

The function $F(\xi)$ is called the Fourier transform of the function $f(x)$. The coordinates x and ξ are inverse to one another, in the sense that if $\Delta x = \lambda$ defines the wavelength of a particle (according to the dualism between particles and waves), then $\xi = \frac{1}{\lambda}$ defines its spatial frequency [7].

Property 1 [7]: The broader the width of $f(x)$ is, the narrower the width of $F(\xi)$ is, and vice versa.

In quantum mechanics we have [1]:

$$p = \frac{h}{\lambda} = h\xi \Leftrightarrow \xi \sim p \Leftrightarrow F(\xi) \sim F(p)$$

with p as the magnitude of the momentum $\vec{p}$; h as the Planck constant; λ as the wavelength; and ξ the spatial frequency. Hence, we conclude that space is the Fourier transform of momentum, and vice versa. Hence, momentum and space are inverse to one another. This leads, as a consequence of property 1, to the uncertainty relation between momentum and space as postulated by Werner Heisenberg:

$$\Delta p_x \Delta x \approx h$$

Corollary 0.2: Hence, momentum and space can be regarded as complementary by the Fourier methodology, in the sense that they are inter-related by a Fourier transform.

0.2 Angular Momentum and Rotational Space

The same arguments as in 0.1 can be used for the rotational case, by replacing the momentum $\vec{p}$ with the angular momentum $\vec{L}$; and by replacing the spatial coordinates (x, y, z) with the angular coordinates $(\varphi_{yz}, \varphi_{zx}, \varphi_{xy})$. Here, we use the right-hand notation where a rotation φ_{yz} in the yz plane is described by a vector in the x direction. The same arguments as in 0.1 lead, as a consequence of property 1, to the uncertainty relation between angular momentum and rotational space as postulated by Werner Heisenberg:

$$\Delta L_{xy} \Delta\varphi_{xy} \approx h$$

Corollary 0.3: Hence, angular momentum and rotational space can be regarded as complementary by the differential operator methodology, in the sense that they are inverse differential operators; and they can be regarded as complementary by the Fourier methodology, in the sense that they are inter-related by a Fourier transform.

0.3 Energy and Time

In physics there is only one possibility to show the complementarity between energy E and time t: The Fourier methodology. The differential operator methodology is not possible, contrary to momentum and space [8]. The reason is that one can define an energy operator but not a time operator, since time enters into Schrödinger's equation not as an operator but as a parameter [8]. Hence, we cannot derive the uncertainty relation between energy and time $\Delta E \Delta t \approx h$ as postulated by Werner Heisenberg with a differential operator methodology [8].

Corollary 0.4: Hence, energy and time behave differently in physics compared with momentum and space.

The Fourier methodology applies for energy and time in the same manner as for momentum and space. We can define

$$f(t) = \int_{-\infty}^{\infty} F(\nu) e^{i2\pi\nu t}\, d\nu$$

with

$$F(\nu) = \int_{-\infty}^{\infty} f(t) e^{-i2\pi\nu t}\, dt$$

Here, we have replaced the spatial coordinate x with the time coordinate t; and the spatial frequency ξ with the temporal frequency ν. The coordinates t and ν are inverse to one another, in the sense that if $\Delta t = T_p$ defines the period of a particle (according to the dualism between particles and waves), then $\nu = \frac{1}{T_p}$ defines its temporal frequency [7].

In quantum mechanics we have [1]:

$$E = h\nu \Leftrightarrow \nu \sim E \Leftrightarrow F(\nu) \sim F(E)$$

with E as the energy; h as the Planck constant; and ν as the temporal frequency. Hence, we conclude that time is the Fourier transform of energy, and vice versa. Hence, energy and time are inverse to one another. This leads, as a consequence of property 1, to the uncertainty relation between energy and time as postulated by Werner Heisenberg:

$$\Delta E \Delta t \approx h$$

Corollary 0.5: Hence, energy and time can be regarded as complementary by the Fourier methodology, in the sense that they are inter-related by a Fourier transform.

0.4 Information and Causality

As will be shown in point 10 below, information and causality also possess an uncertainty relation:

$$\Delta I \Delta k \approx h$$

The value I depicts information as defined in point 4 below; the value k depicts causality as defined in point 8 below. The description of information and causality uses Petri nets as defined in point 6 below. Petri nets do not allow the definition of derivatives and integrals. Hence, neither the differential operator methodology nor the Fourier methodology can be used to show the complementarity between information and causality.

Corollary 0.6: Hence, information and causality behave differently in physics compared with momentum and space on one hand, and also with energy and time on the other hand.

The complementarity between information and causality can be shown by using the approach of Carl Adam Petri as formulated in [3].

For a given net, the approach of [3] uses a triple (S, T, F) with S the set of state-elements; T the set of transition-elements; and F the flow relation. (See figure 6 for a possible Petri net.)

A net is then [3]:

$$Net(S,T,F) :\Leftrightarrow S \cap T = \emptyset, F \subseteq (S \times T) \cup (T \times S), F \cap F^{-1} = \emptyset,$$
$$dom\, F \cup ran\, F = S \cup T$$

Input and output of S elements can be described by input and output relations [3] which define a connectivity matrix C [3]. It is shown in [3] that we obtain a dual net by interchanging S and T which is equivalent to transposing the connectivity matrix C.

Replacing places S with transitions T and vice versa implies the usage of the $\neg$ operator (the not operator). Indeed, the usage of $\neg$ on S leads to T and the usage of $\neg$ on T leads to S since S and T are the only node types of a net [3]. Hence, S which defines information (see point 4 below) is dual to T which defines causality (see point 8 below), and vice versa. We can therefore summarize:

$$S = \neg T \Leftrightarrow I \sim \neg k$$

and

$$T = \neg S \Leftrightarrow k \sim \neg I$$

Corollary 0.7: Hence, information and causality can be regarded as complementary by the $\neg$ operator applied to the places S and to the transitions T of a Petri net.

Corollary 0.8: Hence, we have shown why momentum and space, energy and time, and information and causality are regarded as complementary, respectively, in definition 1 above describing a system.

Corollary 0.9: We have also shown that momentum and space, energy and time, and information and causality behave differently in physics, respectively. Therefore, the ternary structure of the process in definition 2 above, distinguishing among matter (as defined by momenta and angular momenta), energy, and information, is evident.

III. The Structure of Automation Systems

1. Physical Origin of Forces

Physics knows four forces: The strong force described in the quantum field theory of quantum chromodynamics; the electromagnetic and the weak force described in the quantum filed theory of quantum electrodynamics, and the gravitational force described in the theory of general relativity [1].

The strong force acts between quarks by gluons and between the gluons themselves. The strong force does not diminish in strength when the spatial distance between the quarks is increased, and is confined to a range of 1 to 3 femtometers. A complete description of the strong force is still not possible yet by using quantum chromodynamics [1]. Since the approach of this paper lies on the use of a force which is not confined and which varies in space and time, we will not use the strong force in our investigations.

The electromagnetic and weak forces are unified in the electro-weak theory defined by quantum electrodynamics at very high energies [1]. Since we apply forces at low energies where gauge symmetry is spontaneously broken [1], our approach separates the electromagnetic force from the weak force.

The weak force is responsible for radioactive decay [1] and acts between quarks, between leptons, and between quarks and leptons by neutral Z bosons and charged W bosons. Since said bosons have mass, the weak force is of short range (smaller than the radius of an atom). The weak force does not allow bounded states [1]. Since the approach of this paper lies on the use of a force which is not confined and which allows bounded states, we will not use the weak force in our investigations.

The electromagnetic force acts between quarks, between electrically charged leptons, and between quarks and electrically charged leptons by neutral photons [1]. The electromagnetic force manifests itself in the interaction of electrically charged particles, electric and magnetic fields, and electromagnetic waves [1]. The electromagnetic force has infinite range, its effect becoming weaker with increasing distance [1]. Electromagnetic waves are generated by accelerated electrically charged particles and by changing electric and magnetic fields; the electromagnetic waves propagate at the speed of light outward from their source [1].

An electromagnetic force can be created by an electrically charged particle in an electric field $\vec{E}$ (not to be confused with the energy E which is a scalar function described below) and/or by a moving electrically charged particle in a magnetic field $\vec{H}$. The resulting Lorentz force $\vec{F}_L$ is [1]:

$$\vec{F}_L = q\vec{E} + q\vec{v} \times \mu_0\mu_r\vec{H}$$

with q as electric charge, $\vec{v}$ as particle velocity, $\times$ as cross product, μ_0 as vacuum permeability, and μ_r as relative permeability of the surrounding material. We will use the force $\vec{F}_L$ below as one possibility to derive torques, momenta, angular momenta, energy, and information by using electromagnetic values V, A, s (sometimes complemented by the units J, W).

The gravitational force acts between all particles and bodies with a mass. Since energy and mass are equivalent, all forms of energy cause gravitation [1]. Like the electromagnetic force, the gravitational force has infinite range, its effect becoming weaker with increasing distance [1].

Gravitation is described in the theory of general relativity by the curvature of spacetime [1]. Gravitational waves are generated by ripples in the curvature of spacetime, and propagate at the speed of light outward from their source [1]. In the approach of this paper we do not use masses which considerably curve spacetime, like black holes, since this has implications on the coupling of general relativity with quantum field theory, a problem which is not solved yet [1].

Instead we use Newton's approach to gravitation which defines the gravitational force as:

$$\vec{F}_G = G\frac{m_1 m_2}{r^2}\vec{r}_{12}$$

m_1 describing the mass of the first particle/body; m_2 describing the mass of the second particle/body; r describing the radius between the first and second particle/body; $\vec{r}_{12}$ describing a dimensionless unit vector in the direction of the line connecting m_1 and m_2; and G describing the gravitational constant [1]. The Newtonian approach is a good approximation for gravitation when giant masses are avoided. We will use the force $\vec{F}_G$ below as one possibility to derive torques, momenta, angular momenta, energy, and information by using gravitational values kg, m, s (sometimes complemented by the units J, N).

In Newtonian mechanics we have the second law of motion, describing a mechanical force as:

$$\vec{F}_m = m\vec{a}$$

m describing the mass of a body, and $\vec{a}$ its acceleration [1]. We will use the force $\vec{F}_m$ below as one possibility to derive torques, momenta, angular momenta, energy, and information by using mechanical values (which are identical with the gravitational values mentioned above).

2. Force F and Torque T

2.1 Force F

In all three manifestations $(\vec{F}_L, \vec{F}_G, \vec{F}_m)$, the force $\vec{F}$ is, generally spoken, a vector function of space and time [1]. We have:

$$\vec{F}(\vec{r}, t)$$

$\vec{F}$ describing the magnitude and direction of the force vector, $\vec{r}$ describing a vector in three-dimensional space (x, y, z), and t describing a point in time.

The force $\vec{F}$ can be applied in a jump manner to a particle or a body, see figure 1. Hence, the force function does not allow a conservation law.

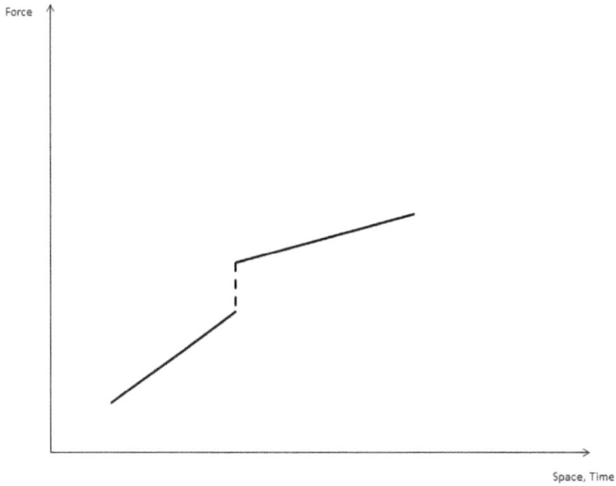

Figure 1: Force in Space and Time

The force $\vec{F}$ is defined by the units $[\vec{F}] = N = \frac{kg\ m}{s^2}$, according to the second law of Newton's laws of motion [1].

2.2 Torque T

The torque $\vec{T}(\vec{\varphi},t) = \vec{r} \times \vec{F}(\vec{r},t)$, (with $\vec{F}(\vec{r},t)$ as force, $\times$ as cross product, and $\vec{r}$ as position vector for the force) [1] is, generally spoken, a vector function of angle $\vec{\varphi}$ and time t. We have:

$$\vec{T}(\vec{\varphi},t)$$

$\vec{T}$ describing the magnitude and direction of the torque vector, $\vec{\varphi}$ describing an angle vector in three-dimensional space (x,y,z), and t describing a point in time.

The torque $\vec{T}$ can be applied in a jump manner to a particle or a body, see figure 2. Hence, the torque function does not allow a conservation law.

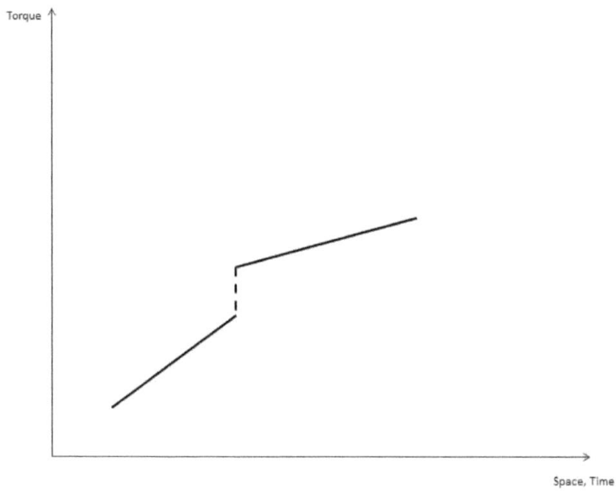

Figure 2: Torque in Space and Time

The torque $\vec{T}$ is hence defined by the units $[\vec{T}] = N\,m = \frac{kg\,m^2}{s^2}$.

3. Momentum p, Angular Momentum L, Overall Momentum, and Energy E

3.1 Momentum p

We can integrate the force in time in order to obtain the momentum [1]:

$$\int_{t_1}^{t_2} \vec{F}(\vec{r},t)dt = \vec{p}(\vec{r})$$

The momentum $\vec{p}$ is a vector function which is dependent of space $\vec{r}$ but not longer of time t, since the time dependence has disappeared by integrating the force $\vec{F}$ in the time interval $\Delta t = t_2 - t_1$.

Due to the integration of a jump in the force function, the momentum function has no jump any more, see figure 3. Hence, the momentum function allows a conservation law.

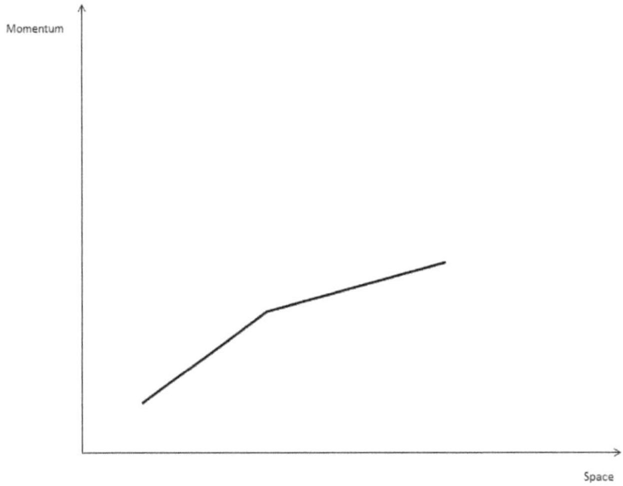

Figure 3: Momentum in Space

One fundamental principle of physics is that the momentum $\vec{p}$ is conserved in a closed system. This can be derived from Newton's laws of motion [1].

Hence, in a closed system we have [1]:

$$\vec{p}_1 = \vec{p}_2 \ (\boldsymbol{Conservation\ of\ momentum})$$

$\vec{p}_1$ describing the momentum before the interaction, and $\vec{p}_2$ after the interaction.

The momentum $\vec{p}$ is hence defined by the units $[\vec{p}] = N\,s = \frac{kg\,m}{s}$.

3.2 Angular Momentum L

When we turn from translational to rotational movements, 5 units have to be mapped [1], see table 1.

Translational Unit	Rotational Unit
$Length\ \vec{l}$ $[\vec{l}] = m$	$Rotation\ Angle\ \vec{\varphi}$ $[\vec{\varphi}] = rad, hence\ dimensionless$
$Velocity\ \vec{v} = \frac{\overrightarrow{dl}}{dt}$ $[\vec{v}] = \frac{m}{s}$	$Angular\ velocity\ \vec{\omega} = \frac{\overrightarrow{d\varphi}}{dt}$ $[\vec{\omega}] = \frac{1}{s}$
$Acceleration\ \vec{a} = \frac{\overrightarrow{dv}}{dt}$ $[\vec{a}] = \frac{m}{s^2}$	$Angular\ acceleration\ \vec{\alpha} = \frac{\overrightarrow{d\omega}}{dt}$ $[\vec{\alpha}] = \frac{1}{s^2}$
$Force\ \vec{F}$ $[\vec{F}] = N$	$Torque\ \vec{T} = \vec{r} \times \vec{F}$ $[\vec{T}] = N\, m$
$Mass\ m$ $[m] = kg$	$Moment\ of\ Inertia\ M = \int_{m_s}^{m_e} r^2 dm$ $[M] = m^2\, kg$

Table 1: Translational and Rotational Units ($\vec{r}$ defines the position vector of the force $\vec{F}$; m_s *and* m_e define the starting and ending points, respectively, when integrating the mass, m, of the body; r defines the distance of the body from the rotation axis.)

We can integrate the torque in time in order to obtain the angular momentum [1]:

$$\int_{t_1}^{t_2} \vec{T}(\vec{\varphi}, t) dt = \vec{L}(\vec{\varphi})$$

The angular momentum $\vec{L}$ is a vector function which is dependent of the angle $\vec{\varphi}$ but not longer of time t, since the time dependence has disappeared by integrating the torque $\vec{T}$ in the time interval $\Delta t = t_2 - t_1$.

Due to the integration of a jump in the torque function, the angular momentum function has no jump any more, see figure 4. Hence, the angular momentum function allows a conservation law.

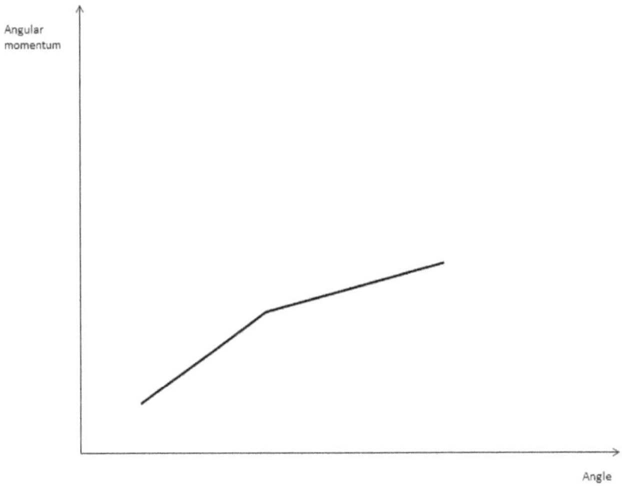

Figure 4: Angular Momentum in the Angle $\vec{\varphi}$

One fundamental principle of physics is that the angular momentum $\vec{L}$ is conserved in a closed system. This can be derived from Newton's laws of motion [1].

Hence, in a closed system we have [1]:

$$\vec{L}_1 = \vec{L}_2 \ (\boldsymbol{Conservation\ of angular\ momentum})$$

$\vec{L}_1$ describing the angular momentum before the interaction, and $\vec{L}_2$ after the interaction.

The angular momentum $\vec{L}$ is hence defined by the units $[\vec{L}] = N\ m\ s = \frac{kg\ m^2}{s}$.

3.3 Overall Momentum

The momentum $\vec{p}$ describes the translational movements; the angular momentum $\vec{L}$ describes the rotational movements. Said two momenta superpose to the overall momentum [1].

3.4 Energy E

We can integrate the force in space and the torque in the angle $\vec{\varphi}$ in order to obtain the energy [1]:

$$\int_{\vec{r}_1}^{\vec{r}_2} \vec{F}(\vec{r},t)d\vec{r} + \int_{\vec{\varphi}_1}^{\vec{\varphi}_2} \vec{T}(\vec{\varphi},t)d\vec{\varphi} = E(t)$$

The energy E is a scalar function which is dependent of time t but not longer of space $\vec{r}$ and on the angle $\vec{\varphi}$, since the space dependence has disappeared by integrating the force $\vec{F}$ in the space interval $\Delta\vec{r} = \vec{r}_2 - \vec{r}_1$, and since the dependence on the angle $\vec{\varphi}$ has disappeared by integrating the torque $\vec{T}$ in the angle interval $\Delta\vec{\varphi} = \vec{\varphi}_2 - \vec{\varphi}_1$. Due to the integration of a jump in the force and torque functions, the energy function has no jump any more, see figure 5. Hence, the energy function allows a conservation law.

One fundamental principle of physics is that the energy E is conserved [1]. Hence, we always have

$$E_1 = E_2 \ (\boldsymbol{Conservation\ of\ energy})$$

E_1 describing the energy before the interaction, and E_2 after the interaction.

The energy E is defined by the units $[E] = J = N\,m = \frac{kg\,m^2}{s^2}$.

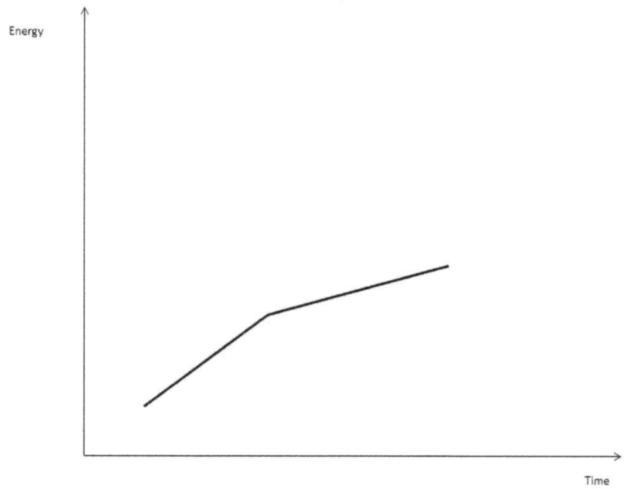

Figure 5: Energy in Time

4. Information I

We can integrate the force in space and time and the torque in the angle $\vec{\varphi}$ and in time, or we can integrate the momentum in space and the angular momentum in the angle $\vec{\varphi}$, or we can integrate the energy in time to obtain the information I [4]:

$$\iint_{t_1\vec{r}_1}^{t_2\vec{r}_2} \vec{F}(\vec{r},t)d\vec{r}dt + \iint_{t_1\vec{\varphi}_1}^{t_2\vec{\varphi}_2} \vec{T}(\vec{\varphi},t)d\vec{\varphi}dt =$$
$$\int_{\vec{r}_1}^{\vec{r}_2} \vec{p}(\vec{r})d\vec{r} + \int_{\vec{\varphi}_1}^{\vec{\varphi}_2} \vec{L}(\vec{\varphi})d\vec{\varphi} =$$
$$\int_{t_1}^{t_2} E(t)dt =$$
$$I$$

The information I is a scalar which is neither dependent on space nor on time, since the space dependence has disappeared by integrating the force $\vec{F}$ in the space interval $\Delta\vec{r} = \vec{r}_2 - \vec{r}_1$ and the torque $\vec{T}$ in the angle interval $\Delta\vec{\varphi} = \vec{\varphi}_2 - \vec{\varphi}_1$, and since the time dependence has disappeared by integrating the force $\vec{F}$ and the torque $\vec{T}$ in the time interval $\Delta t = t_2 - t_1$.

Hence, information I is described by a scalar (by a number) and not by a function, like the force, momentum, angular momentum, or energy [4].

I summarizes the momentum of the space interval $\Delta\vec{r} = \vec{r}_2 - \vec{r}_1$, the angular momentum of the angle interval $\Delta\vec{\varphi} = \vec{\varphi}_2 - \vec{\varphi}_1$, and the energy of the time interval $\Delta t = t_2 - t_1$ in an index I.

The information I is defined by the units [4]:

$$[I] = J\,s = V\,A\,s^2 = W\,s^2 = N\,m\,s = \frac{kg\,m^2}{s} := bit.$$

Corollary 1: Hence, we have defined the information I in a physical manner by using a force and torque applied to a particle/body, and have connected this information to bits used in information technology [4].

Corollary 2: Since information is not defined as a function of space and/or time but by a number, we cannot formulate a law of conservation of information at this stage, since we cannot speak of time or space before and after the interaction. Instead, I summarizes the momentum, the angular momentum, and the energy of the interaction process and represents the result as a number defining bits [4].

Picture 1 summarizes the relationship among force, torque, momentum, angular momentum, energy, and information.

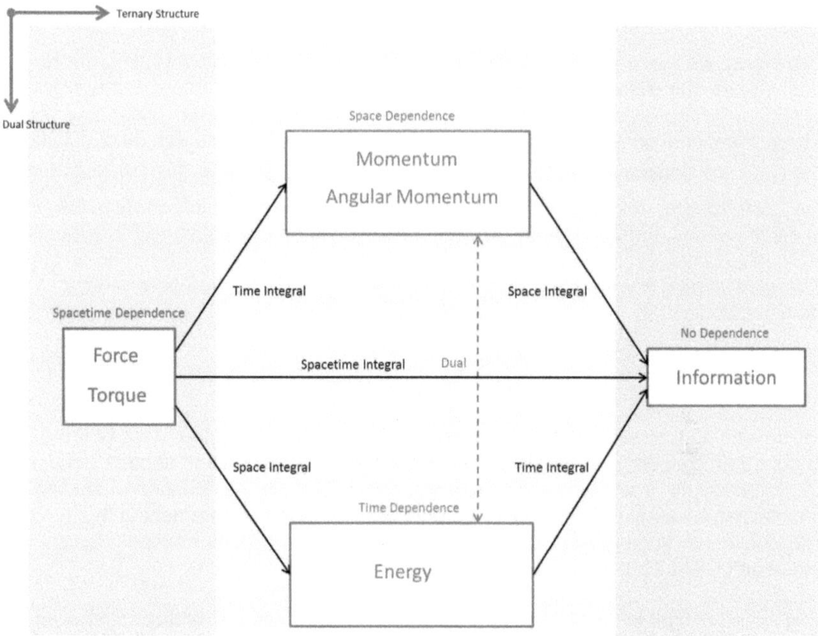

Picture 1: Relationship among force, torque, momentum, angular momentum, energy, and information

5. Planck Units

In physics, we can use five universal constants (the Planck constants) in order to define units of measurement [2].

The gravitational constant, G; the speed of light in vacuum, c; the Planck constant, h; the Coulomb constant, $\frac{1}{4\pi\varepsilon_0}$; the Boltzmann constant k_B. Since we are interested in the force, torque, momentum, angular momentum, energy, and information, but not in electrical and thermodynamic processes, we focus on the first three Planck units.

The gravitational constant, G, is important when defining the force between two particles/bodies [1]:

$$\vec{F}_G = G\frac{m_1 m_2}{r^2}\vec{r}_{12} \quad (1)\text{: } Matter\ equation$$

m_1 describing the mass of the first particle/body; m_2 describing the mass of the second particle/body; r describing the radius between the first and second particle/body; $\vec{r}_{12}$ describing a dimensionless unit vector in the direction of the line connecting m_1 and m_2. The Force applied to matter is the key element in the introduction of momentum, angular momentum, energy, and information. Hence, equation (1) shall be defined [4] the **matter equation**.

The speed of light in vacuum, c, is important when defining the relationship between matter (given by the mass m_0 of the resting particle) and energy E_0 of the resting particle [1]:

$$E_0 = m_0 c^2 \quad (2)\text{: } Energy\ equation$$

Hence, equation (2) shall be defined [4] the **energy equation**.

The Planck constant, h, is important when defining the relationship between energy, E, and frequency/time [1]:

$$E = hf \quad (3)\text{: } Information\ equation$$

E defines the energy of the particle/wave and f defines its frequency. We have $f = \frac{1}{T}$, with T being the period of the wave. The Planck constant, h, has the same dimension as information: $[I] = J\ s = V\ A\ s^2 = W\ s^2 = N\ m\ s = \frac{kg\ m^2}{s} := bit$. Indeed, the equations

$$\Delta p_x \Delta x \approx h \quad and \quad \Delta E \Delta t \approx h$$

known as Heisenberg's uncertainty principle [1] state that the information defined by $\Delta p_x \Delta x$ or $\Delta E \Delta t$ is at least h. Hence, h is the smallest possible information. Hence, we shall call equation (3) as [4] the **information equation**.

6. Petri Nets

As already mentioned in corollary 2, I summarizes the momentum, angular momentum, and the energy of the interaction process and represents the result as a number defining bits. This fact can be easily depicted by Petri nets. Petri nets are defined in [3]. A possible Petri net is shown in figure 6.

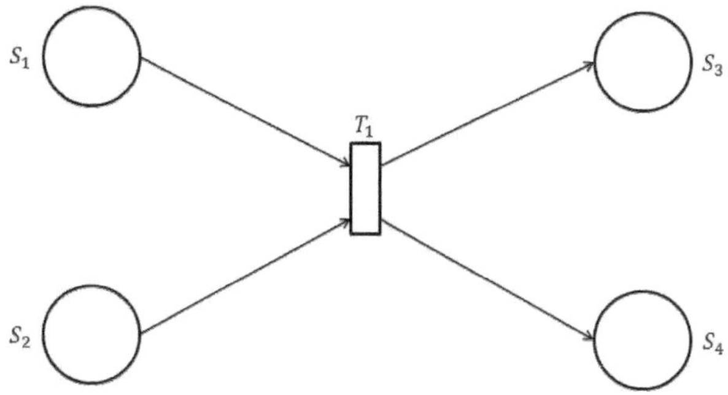

Figure 6: Petri Net

The circles $S_1 - S_4$ are called places and represent the state of a system. We can identify the places with information as described above. Hence, the places $S_1 - S_4$ identify four information pieces $I_1 - I_4$ [4].

The rectangle T_1 is called transition and represents, together with the arrows, the causal relationship between the Information pieces.

Corollary 3: Hence, Petri nets are causal nets representing the causal relationship of information pieces [4].

Corollary 4: Information as defined above cannot be shown in space and/or time, since said information is not a function of space and/or time. But said information can be shown in a causal net [4].

7. Physical Measurements and Conservation Laws

It is a well known principle of quantum theory that a measurement disturbs the system which is measured [1]. Such an example is shown in figure 7 [4].

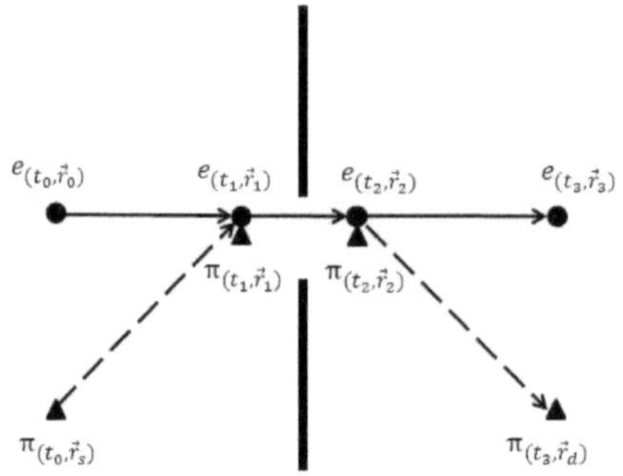

Figure 7: Physical Measurement

The triangle represents a photon π used to perform a measurement, and the circle represents an electron e which is measured by the photon π. The photon starts at time t_0, arrives at time t_1 at the electron, interacts with the electron from time t_1 to time t_2, is scattered from the electron at time t_2, and arrives at the observer at time t_3. The same applies for the spatial coordinates r_i, of course, and needs not to be repeated [4].

The interaction between electron and photon is guided, of course, by the basic laws of physics like the law of conservation of energy, and the law of conservation of momentum and angular momentum (see the discussion above).

We have before the interaction [4], hence before the time t_1:

$$E_{bf} = E_1 + E_2$$

E_1 describing the energy of the photon before the interaction, and E_2 describing the energy of the electron before the interaction.

We have after the interaction [4], hence after the time t_2:

$$E_{af} = E_3 + E_4$$

E_3 describing the energy of the photon after the interaction, and E_4 describing the energy of the electron after the interaction.

According to the law of conservation of energy we have:

$$E_1 + E_2 = E_3 + E_4 \;\; (*)$$

The energy $E_1 + E_2$ is transformed during the interaction (hence between t_1 and t_2) by an energy flow to the energy $E_3 + E_4$; and the energy $E_3 + E_4$ is set up during the interaction (hence between t_1 and t_2) by said energy flow from the energy $E_1 + E_2$.

Hence, we can integrate equation $(*)$ in time for the time period of interaction (from t_1 to t_2), and obtain [4]:

$$\int_{t_1}^{t_2} E_1 dt + \int_{t_1}^{t_2} E_2 dt = \int_{t_1}^{t_2} E_3 dt + \int_{t_1}^{t_2} E_4 dt \quad (**)$$

Equation $(**)$ can be written as [4]:

$$I_1 + I_2 = I_3 + I_4 \ (\boldsymbol{Conservation\ of\ information})$$

Remark: We could have derived the law of conservation of information also by integrating the momentum in space and the angular momentum in the angle $\vec{\varphi}$, and by using the law of conservation of momentum and angular momentum. We could also have derived the law of conservation of information by integrating the force in space and time and the torque in the angle $\vec{\varphi}$ and in time, and by using the law of conservation of momentum, angular momentum, and energy [4].

Corollary 5: The validity of the laws for conservation of energy, momentum, and angular momentum leads to the law of conservation of information. No information gets lost during a measurement [4].

Hence, we have four conservation laws [4]:

$$\vec{p}_1 = \vec{p}_2 \ (\boldsymbol{Conservation\ of\ momentum})$$

$$\vec{L}_1 = \vec{L}_2 \ (\boldsymbol{Conservation\ of\ angular\ momentum})$$

$$E_1 = E_2 \ (\boldsymbol{Conservation\ of\ energy})$$

$$I_1 = I_2 \ (\boldsymbol{Conservation\ of\ information})$$

Corollary 6: Whereas the laws of conservation of energy, momentum, and angular momentum can be directly observed in the local reference frame of the interacting particles, the law of conservation of information can only be observed during a measurement by using the local reference frame of the particles and the local reference frame of the observer [4].

Corollary 7: According to Albert Einsteins's special theory of relativity the reference frames of particles and observer are connected by a Lorentz transformation [1]. Hence, space and time get transformed from one reference frame to another, and the momentum, angular momentum, and energy also get transformed between both reference frames [1]. Due to $(*)$ and $(**)$ the information I gets transformed between both reference frames connected by a Lorentz transformation [4]. It will be shown in corollary 9 below that I is an invariant with respect to Lorentz transformations.

Corollary 8: Information cannot be observed in a spacetime diagram like momentum, angular momentum, and energy, since I is not a function of spacetime. But information can be observed in a causal net. Since the causal net consists of two dimensions (places and transitions) [3], the observation of information adds two new dimensions [4].

Contrary to Newtonian mechanics which does not allow gravitational waves, electromagnetic theory allows electromagnetic waves [1]. We have the following equation showing the propagation of an electromagnetic wave $\vec{\Psi}_{EM}$ [1]:

$$\vec{\Psi}_{EM}(\vec{r},t) = A \cdot e^{i(\vec{k}\vec{r}-\omega t)}$$

with $\vec{r}$ as the position vector of the wave in three dimensional space, t as the time coordinate of the wave, A as the amplitude of the wave, i as the imaginary unit ($i^2 = -1$) of the complex numbers ($\mathbb{C}$), $\vec{k}$ as the wave vector in three dimensional space (not to be confused with the causality axis k described below), $\vec{k}$ showing in the direction of the propagation of the wave, and ω as the angular frequency of the wave.

The wave vector $\vec{k}$ defines a measure for the momentum and the angular momentum of the wave [1]. The angular frequency ω defines a measure for the energy of the wave [1].

Corollary 5.1: Hence, we conclude that the electromagnetic wave carries momentum, angular momentum, and energy during the wave propagation process.

Corollary 6.1: We further conclude that the electromagnetic wave does not carry information during the wave propagation process, since information requires the definition of a spatial range $\Delta\vec{r} = \vec{r}_2 - \vec{r}_1$, of an angular range $\Delta\vec{\varphi} = \vec{\varphi}_2 - \vec{\varphi}_1$, and of a temporal range $\Delta t = t_2 - t_1$ which are not present in the equation of the propagation of the wave.

In order to use the spatial, angular, and temporal ranges of corollary 6.1, measurements must be defined, such that the wave $\vec{\Psi}_{EM}$ interacts with measuring waves/particles in said spatial, angular, and temporal ranges.

Corollary 6.2: Without measurement, the propagation of electromagnetic waves carries momentum, angular momentum, and energy but no information. With measurement in a spatial range $\Delta\vec{r} = \vec{r}_2 - \vec{r}_1$, in an angular range $\Delta\vec{\varphi} = \vec{\varphi}_2 - \vec{\varphi}_1$, and in a temporal range $\Delta t = t_2 - t_1$, the electromagnetic wave additionally carries information.

8. Information and Causality

As already discussed, information is naturally depicted by causal nets. A causal net possesses two distinct elements: the places (represented by circles) and the transitions (represented by rectangles). Figure 8 shows the two possibilities of an elementary causal net [3].

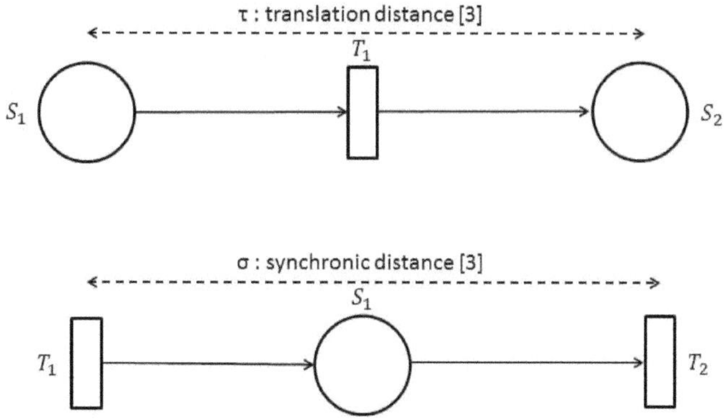

Figure 8: The two Elementary Structures of a Causal Net

As shown in [3], every place S has to be connected to a transition T but not to another place; and every transition T has to be connected to a place S but not to another transition. Therefore, the two possibilities $S_1 - T_1 - S_2$ or $T_1 - S_1 - T_2$ define the elementary net structures. It is proven in [3] that the places S and the transitions T are dual entities, and that a causal net defines a continuum.

Carl Adam Petri shows in [3] that one can define a translation distance τ between the places, and a synchronic distance σ between the transitions. The longer the distances τ and σ are, the more places and transitions, respectively, are crossed. Hence, τ and σ define two possible axes, one for the places and one for the transitions, respectively. In [3] τ and σ are dimensionless, since they are applied to the abstract structure of causal nets.

We have already shown in corollary 3 that we can identify the places with information pieces I_i. In this case, the translation distance τ defines an information axis, I, defining distances among the information pieces I_i, bearing in mind that several information pieces can occupy the same location on the information axis defined by τ. In the case of information, τ is not longer dimensionless (like for abstract causal nets), but possesses the dimension of information, namely $[I] = J\, s = V\, A\, s^2 = W\, s^2 = N\, m\, s = \frac{kg\, m^2}{s} := bit.$

The transitions define the interaction of the information pieces I_i according to equation $(**)$ above. Hence, we have information pieces I_{S1} ... I_{Sn} before the interaction (S standing for "source"), such that said information pieces are connected by the transition to information pieces $I_{D1} \dots I_{Dm}$ after the interaction (D standing for "destination"), and such that $\sum_{i=1}^{i=n} I_{Si} = \sum_{i=1}^{i=m} I_{Di}$ due to the law of conservation of information [4].

Otherwise stated, we have a vector of information pieces before the interaction [4]: $\vec{I_S} = \begin{pmatrix} I_{S1} \\ \vdots \\ I_{Sn} \end{pmatrix}$, and we have a vector of information pieces after the interaction [4]: $\vec{I_D} = \begin{pmatrix} I_{D1} \\ \vdots \\ I_{Dm} \end{pmatrix}$. The transition acts as a matrix [4] $\underline{K}$ projecting $\vec{I_S}$ to $\vec{I_D}$. The matrix $\underline{K}$ has hence the structure $\underline{K} = \begin{pmatrix} k_{11} & \cdots & k_{1n} \\ \vdots & \ddots & \vdots \\ k_{m1} & \dots & k_{mn} \end{pmatrix}$.

Hence, we have the equation [4]

$$\begin{pmatrix} I_{D1} \\ \vdots \\ I_{Dm} \end{pmatrix} = \begin{pmatrix} k_{11} & \cdots & k_{1n} \\ \vdots & \ddots & \vdots \\ k_{m1} & \dots & k_{mn} \end{pmatrix} \cdot \begin{pmatrix} I_{S1} \\ \vdots \\ I_{Sn} \end{pmatrix}$$

showing how the information before the interaction is projected on the information after the interaction. Hence, the matrix $\underline{K}$ showing the interaction described by a transition is dimensionless. We identify the matrix $\underline{K}$ with causality, and conclude that causality is dimensionless, contrary to information [4].

We can use the synchronic distance σ to define a causality axis, k, defining distances among the matrices $\underline{K_i}$, bearing in mind that several matrices can occupy the same location on the causality axis defined by σ. In the case of causality, σ remains dimensionless like for abstract causal nets.

The dimensions for information I and for causality k are the two additional dimensions mentioned in corollary 8 [4].

9. Special Relativistic Effects on Space, Time, Momenta, Angular Momenta, Energy, Information, and Causality

In special relativity space and time are transformed from a resting frame to a frame moving with velocity $\vec{v}$ [1]. The magnitude of $\vec{v}$ shall be called v. The length l' measured by the moving observer is contracted with respect to the length l_0 measured by the resting observer according to the equation [1]:

$$l' = \sqrt[2]{1 - \left(\frac{v^2}{c^2}\right)}\, l_0 = \frac{1}{\gamma} l_0$$

with c defining the speed of light in vacuum. γ is called the Lorentz factor [1].

The angle φ' measured by the moving observer is contracted with respect to the angle φ_0 measured by the resting observer according to the equation [1]:

$$\varphi' = \sqrt[2]{1 - \left(\frac{v^2}{c^2}\right)}\varphi_0 = \frac{1}{\gamma}\varphi_0.$$

The time t' measured by the moving observer is dilated with respect to the time t_0 measured by the resting observer according to the equation [1]:

$$t' = \sqrt[2]{1 - \left(\frac{v^2}{c^2}\right)}t_0 = \frac{1}{\gamma}t_0.$$

When v approaches c, the length l' and the angle φ' as measured by the moving observer tend towards 0, and the time flow t' measured by the moving observer tends towards 0 (otherwise stated: time flow stops for the moving observer when $v = c$).

The relativistic momentum p' is given by the equation [1]:

$$p' = \gamma m_0 v = \frac{m_0 v}{\sqrt[2]{1 - \frac{v^2}{c^2}}}$$

with m_0 as the mass of the resting particle. (We focus here on the magnitudes of the vectors, since the vector directions are not relevant for our investigations.)

The relativistic angular momentum L' can be derived as follows. We set up the cross product $\times$ between the spatial position vector $\vec{r}$ and the particle velocity $\vec{v}$ to obtain [1]:

$$L' = \gamma|m_0\vec{r} \times \vec{v}| = \left|\frac{m_0\vec{r} \times \vec{v}}{\sqrt[2]{1 - \frac{v^2}{c^2}}}\right|.$$

(The vector $\vec{r}$ is not subject to Lorentz contraction, since the movement occurs around $\vec{r}$). The relativistic energy E' is given by the equation [1]:

$$E' = \gamma m_0 c^2 = \frac{m_0 c^2}{\sqrt[2]{1 - \frac{v^2}{c^2}}}.$$

Hence, when v approaches c, the particle momentum, angular momentum, and energy tend towards ∞.

Information couples momentum with space, angular momentum with angle, and energy with time in a multiplicative manner as already described in the equation of point 4 above defining the information I. Therefore, multiplications between l' and p', between φ' and L', and between t' and E' lead in the cancellation of the Lorentz factor γ.

Corollary 9: Hence, we conclude that the information I' measured by the moving observer is the same as the information I_0 measured by the resting observer. Information is therefore an invariant with respect to Lorentz transformations.

Point 8 above defines the causality k using the matrix $\underline{K}$. In the reference frame of the moving observer we have then:

$$\begin{pmatrix} I'_{D1} \\ \vdots \\ I'_{Dm} \end{pmatrix} = \begin{pmatrix} k'_{11} & \cdots & k'_{1n} \\ \vdots & \ddots & \vdots \\ k'_{m1} & \cdots & k'_{mn} \end{pmatrix} . \begin{pmatrix} I'_{S1} \\ \vdots \\ I'_{Sn} \end{pmatrix}.$$

Since information is an invariant with respect to Lorentz transformations (see corollary 9 above) the vectors $\begin{pmatrix} I'_{D1} \\ \vdots \\ I'_{Dm} \end{pmatrix}$ and $\begin{pmatrix} I'_{S1} \\ \vdots \\ I'_{Sn} \end{pmatrix}$ of the moving observer are the same as the vectors $\begin{pmatrix} I_{D10} \\ \vdots \\ I_{Dm0} \end{pmatrix}$ and $\begin{pmatrix} I_{S10} \\ \vdots \\ I_{Sn0} \end{pmatrix}$ of the resting observer, respectively.

Therefore, the matrix $\begin{pmatrix} k'_{11} & \cdots & k'_{1n} \\ \vdots & \ddots & \vdots \\ k'_{m1} & \cdots & k'_{mn} \end{pmatrix}$ of the moving observer must be the same as the matrix $\begin{pmatrix} k_{110} & \cdots & k_{1n0} \\ \vdots & \ddots & \vdots \\ k_{m10} & \cdots & k_{mn0} \end{pmatrix}$ of the resting observer.

Corollary 10: Hence, we conclude that the causality k' measured by the moving observer is the same as the causality k_0 measured by the resting observer. Causality is therefore, like information, an invariant with respect to Lorentz transformations.

10. Quantum Mechanical Effects on Space, Time, Momenta, Angular Momenta, Energy, Information, and Causality

In physics, the conservation of momentum, angular momentum and energy are fundamental principles as discussed above. Furthermore, we have derived a conservation of information in this paper.

Corollary 11: Since momentum and angular momentum are quantities describing the features of particles, momentum and angular momentum can be viewed as features of matter. Therefore, we have conservation laws describing how matter, energy, and information are conserved [4].

In quantum mechanics, Werner Heisenberg has shown that the uncertainty principle is valid when measuring the momentum range Δp_x and spatial range Δx of a particle, or the energy range ΔE and the temporal location range Δt of a particle. The same applies, of course, for the measurement of the angular momentum range ΔL_{xy} of a particle and the angle location range $\Delta\varphi_{xy}$ of said particle. Carl Adam Petri shows in [3] that the causal nets also possess an uncertainty relation. We therefore conclude that the measurement of the information range ΔI and the measurement of the causality range Δk are also uncertain. Since the information range ΔI and the causality range Δk always overlap (due to the structure of the causal nets), we always have to consider ΔI and Δk in combination. Therefore, we have [4]:

$$\Delta p_x \Delta x \approx h \ \ (\boldsymbol{Uncertainty\ of\ matter\ in\ space})$$

$$\Delta L_{xy} \Delta\varphi_{xy} \approx h \ \ (\boldsymbol{Uncertainty\ of\ rotating\ matter\ in\ rotational\ space})$$

$$\Delta E \Delta t \approx h \ \ (\boldsymbol{Uncertainity\ of\ energy\ in\ time})$$

$$\Delta I \Delta k \approx h \ \ (\boldsymbol{Uncertainty\ of\ information\ in\ causality})$$

The uncertainty of information in causality can be formally derived as follows. h is the smallest possible information (see the last paragraph of point 5 above). Hence, ΔI cannot be smaller than h, leading to the quantization of information. Hence, Δk has to be the identity element in this case, meaning that the particle is in a stable state, not interacting with other particles [4].

The first uncertainty relation is explained by Werner Heisenberg in the following manner [4]. The more accurate the measurement of a spatial position of a particle, the smaller the wavelength of the measuring wave must be. But the smaller the wavelength of the measuring wave is, the bigger the momentum of the measuring wave is, such that the impact on the momentum of the measured particle is big, leading to a big uncertainty of said momentum. The more accurate the measuring of a momentum of a particle, the bigger the wavelength of the measuring wave must be, in order not to influence the particle momentum. But the bigger the wavelength of the measuring wave is, the less precise the measurement of the spatial position of the particle is, leading to a big uncertainty is said position. Hence, momentum and space cannot be determined with big accuracy at the same time.

The second uncertainty relation can be explained in a corresponding manner [4]. The more accurate the measurement of a rotation angle of a particle, the smaller the wavelength parallel to the rotational movement of the measuring wave must be. But the smaller the wavelength parallel to the rotational movement of the measuring wave is, the bigger the angular momentum of the measuring wave is, such that the impact on the angular momentum of the measured particle is big, leading to a big uncertainty of said angular momentum. The more accurate the measuring of an angular momentum of a particle, the bigger the wavelength parallel to the rotational movement of the measuring wave must be, in order not to influence the particle angular momentum. But the bigger the wavelength parallel to the rotational movement of the measuring wave is, the less precise the measurement of the rotation angle of the particle is, leading to a big uncertainty is said rotation angle. Hence, angular momentum and rotation angle cannot be determined with big accuracy at the same time.

The third uncertainty relation is explained by Werner Heisenberg in a similar manner [4]. The more accurate the measurement of a temporal position of a particle, the bigger the frequency of the measuring wave must be. But the bigger the frequency of the measuring wave is, the bigger the energy of the measuring wave is, such that the impact on the energy of the measured particle is big, leading to a big uncertainty of said energy. The more accurate the measuring of an energy of a particle, the smaller the frequency of the measuring wave must be, in order not to influence the particle energy. But the smaller the frequency of the measuring wave is, the less precise the measurement of the temporal position of the particle is, leading to a big uncertainty is said position. Hence, energy and time cannot be determined with big accuracy at the same time.

The fourth uncertainty relation can be explained as follows [4]. The more accurate the measurement of the information of a particle, the more isolated the particle must be, in order to be able to measure its energy, momentum, and angular momentum states. But the more isolated said particle is, the more interactions with other particles are destroyed, leading to a big uncertainty of causality. The more accurate the measurement of causality shall be, the more particle interactions shall be observed. But the more particle interactions shall be observed, the less isolated single particles shall be, leading to a big uncertainty of the information of a single particle. Hence, information and causality cannot be determined with big accuracy at the same time.

Corollary 12: The substances matter (as defined by its momentum and angular momentum), energy, and information are quantized and lead to an uncertainty relation in their existence forms, space, time, and causality [4].

Corollary 13: In the physical view of a single local reference frame, there are 8 dimensions: 3 dimensions of translational and rotational space, $\left((x,\varphi_{yz}),(y,\varphi_{zx}),(z,\varphi_{xy})\right)$, 3 complementary dimensions of space given by the overall momenta $\left((p_x,L_{yz}),(p_y,L_{zx}),(p_z,L_{xy})\right)$ [1], 1 dimension of time (t), and 1 complementary dimension of time given by the energy (E) [1]. In the physical view of several local reference frames, there are 10 dimensions: the 8 dimensions of the single local reference frame, 1 dimension of causality (k), and 1 complementary dimension of causality given by the information (I) [4].

11. Final Remarks on the Nature of Information

The derivation of information achieved in this paper, and especially the derivation of the law of conservation of information, appear astonishing viewed from the point of view of information technology. Indeed, in information technology we have Shannon's definition of information by using probabilities. Furthermore, information can be always destroyed and created, such that a law of conservation of information makes no sense in information technology [4].

This apparent discrepancy can be best resolved by analysing the nature of energy in physics and engineering. The law of conservation of energy is a very old principle in physics, and is the basis for many physical derivations. Nevertheless, when performing engineering, like constructing an electric power station, one arrives at the conclusion that not all energy forms are suitable for the set task. Indeed, when converting the mechanical energy of the turbine into electrical energy by the generator, one obtains the equation [4]:

$$E_{mechanical} = E_{electrical} + E_{heat}$$

This equation depicts the thermodynamic principle that the mechanical energy of the turbine cannot be entirely transformed to electrical energy; instead heat is produced during the energy transformation process [1].

Physically spoken, no energy is lost, since the energy before the interaction (the mechanical energy) is exactly the same as the energy after the interaction (the electrical and the heat energy). But from the point of view of an engineer, there is a loss, since the heat energy cannot be used and therefore defines a technological loss. Hence, in engineering, the heat energy is minimized as much as possible. For this reason, an energy efficiency factor η_e is defined in engineering [4]:

$$\eta_e = \frac{E_{electrical}}{E_{mechanical}}$$

This factor is between 0 and 1, and shall be as much as possible near 1, but can never achieve 1 due to thermodynamic principles [1]. This shows that the fundamental physical principle of the conservation of energy is not always of interest in engineering; instead physically equivalent energy types are rated in engineering, defining some energy types as desired and others as not desired, such that energy losses might appear in engineering [4].

The same reasoning as for energy applies also for information. Indeed, it has been shown in this paper that information is conserved in physics. Hence, from a physical point of view, information is never lost. But from the point of view of information technology, some information pieces $I_1 \dots I_i$ might be desired, since they represent information which can be used for a given purpose, whereas other information pieces $I_{i+1} \dots I_n$ might be undesired, since they cannot be used for said given purpose [4].

Hence, information pieces are rated in information technology with respect to their usability. Therefore, in information technology an information loss is possible, due to the fact of the presence of desired and undesired information. One could define an information efficiency factor η_i for information technology [4]:

$$\eta_i = \frac{I_{usable}}{I_{overall}}$$

This factor is between 0 and 1, and shall be as much as possible near 1, but can never achieve 1 due to uncertainty principles between information and causality. This shows that the fundamental physical principle of the conservation of information is not always of interest in information technology; instead physically equivalent information pieces are rated in information technology, defining some information pieces as desired and others as not desired, such that information losses might appear in information technology [4].

Shannon's definition of information by using probabilities is then another form of defining desired and not desired information by weighting some information pieces more than others. This is in line with the point of view of engineering which rates energy, information, and momenta (although momenta have not been addressed in this discussion of point 11). But the ratings performed in engineering do not contradict the physical conservation laws for momentum, angular momentum, energy, and information [4].

Thus we have derived a well founded reasoning for definition 1 defining a system in automation systems [6].

IV. The Functionality of Automation Processes

Distributed automation systems consist of components for the handling of material flows. These material resources demand on their part energy handling resources. For the influence of the material and energy resources information between the components has to be exchanged; therefore information resources are required [6].

The material, energy and information systems of automation engineering interact with each other. Only an integrated description of the resources respecting all interactions guarantees a correct and exact modelling and prediction of the system behavior [6].

The automation theory is characterized by a given number of operations per spatial volume ΔV and per time interval Δt conducted by the technical resources. This fact implies a continuity equation of the substances matter, energy and information, respectively, in the world of automation engineering. Spatial, temporal and causal translations of the substances can thus be regarded as transformations of coordinates which are invariant in the quadratic form, like in the special theory of relativity [6].

The formulation of a generalized continuity equation including spatial, temporal and causal translations leads to balance equations for the different components of automation systems. Balance equations can be evaluated for the computation of task scheduling algorithms respecting the actual load of the technical resource [6].

For the formulation of the generalized continuity equation two problems have to be solved: first, the measuring of causal translations which on its part implies the second problem of a geometric derivation of causality. In physics the structure of the existence forms space and time is described by spatial and temporal (clocks) measures [6].

The geometric derivation allows the integration of the existence forms with the four dimensional spacetime of the special theory of relativity. This integration respects the physical demands on spacetime translations describing them in a correct and precise form [6].

Petri nets offer the excellent property of permitting a geometric derivation of causality. The causal measures won from Petri nets define the necessary causal ordering relations. A five dimensional generalized spacetime of automation engineering including causality can now be derived [6].

The behavior of automation systems depends on the different causal states and of allowed transitions implied by the selection rules for causal transitions. From this point of view automation systems behave like physical microsystems. Therefore the matrix transition elements of quantum mechanics can be transferred to causal matrix transition elements. These operators are won analyzing the Petri net representation of automation systems [6].

For the realization of spatial, temporal and causal translations technical resources are required. In automation engineering the three classes material, energy and information resources appear and interact with each other. Each class can be divided into three functional basic elements: causal-processing (P), spatial-transportation (T) and temporal-storage (S) functions [5].

In a production system P functions are represented by the machines, T functions by the transportation system, and S functions by the storage units. Energy systems own turbines and rural subscriber lines which act like P and T functions respectively; accumulators represent the S functions. Processors and communication systems are the P and T functions of information systems, respectively; the S functions are specified by the memory of the information system. Table 2 summarizes these reflections [5].

Substance	P	T	S
Matter	Production	Transportation	Storage
Energy	Transformation	Distribution	Storage
Information	Processing	Communication	Storage

Table 2: Substances and Functional Basic Elements

Every functional basic element is afflicted with an evaluation time *t*, which is required for the task. The P functions effect a transformation:

$$P = f(x, t_P)$$

x describes the causal state of the resource, t_P is the processing time. The P function transforms the causal state of the system in the time t_P. For the T functions equation

$$T = f(r, t_T)$$

describes the spatial translations *r* conducted by the T functions in time t_T (transportation time). Equation

$$S = f(t, t_S)$$

specifies the temporal translations *t* conducted by the S functions in time t_S. Parameters t_P, t_T and t_S are determined by the implementation of the resources.

The P, T, and S functions depict physical processes in a natural manner, see table 3.

	Space (T)	Time (S)	Causality (P)
Matter	Magnetic (Inductor L)	Electric (Capacitor C)	Photonic (Resistor R)
Energy	Heat (Transitive)	Internal Energy (Potential)	Work (Interacting)
Information	Transmission System (Communicating)	Storage System (Memorizing)	Processing System (Computing)

Table 3: Physical Processes

Causal dependencies of automation systems can be described in a simple graphic and precise mathematic way using Petri nets, as already explained above. According to figure 8 we have two possible elementary net structures. We can use the translation distance τ between the places, and/or the synchronic distance σ between the transitions. Since the approach using P, T, and S functions is based on measuring the system states when defining a process, we shall use the translation distance τ between the places as our measure for information and causality [5].

Remark 1: Since the translation distance τ and the synchronic distance σ are always interleaved, due to the structure of the Petri nets, and since we do not have to distinguish between information and causality in the functional approach of a process, we can use a single measure, τ or σ. Since our approach lies on causal states, we will use the translation distance τ, and will call τ a measure of causality defining a causal axis of processes.

The measure τ is applied to pure nets in this case, such that τ is dimensionless like for pure nets, and does not possess the dimension of information. Figure 9 and table 4 show how this causal measure can be applied to a random causal structure.

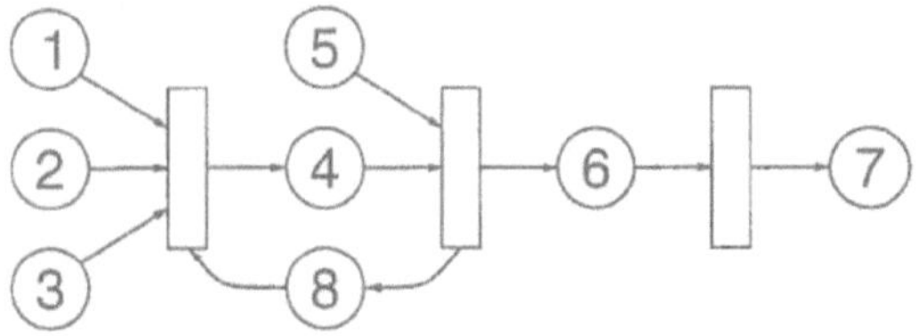

Figure 9: Causal Structure

	1	**2**	**3**	**4**	**5**	**6**	**7**	**8**
1	0	0	1	1	2	3	3	4
2	0	0	1	1	2	3	3	4
3	-1	-1	0	0	1	2	2	3
4	-1	-1	0	0	1	2	2	3
5	-2	-2	-1	-1	0	1	1	2
6	-3	-3	-2	-2	-1	0	-2	1
7	-3	-3	-2	-2	-1	2	0	3
8	-4	-4	-3	-3	-2	-1	-3	0

Table 4: Causal Distances according to τ

The rows represent the starting causal states, the columns represent the target causal states. Row *i* and column *j* indicate the causal distance between state *i* and state *j*. If *i=j* (start and target state are identical) the distance is 0. If there is no other possibility to proceed from a state to the other than to proceed once in direction of the arrows and once in the opposite arrow direction then the causal distance is also set to 0 because the causal states are independent of each other. An example is given in figure 9 between states 1 and 2 of the Petri net [6].

If one can proceed from a state to the other as well as in arrow direction as also against arrow direction, a specification needs to be formulated about the positioning of the coordinate system. This specification explains which of the two alternatives is counted as positive and which as negative. Like in affine geometry the causal distances of opposite direction are counted as negative. Every feedback loop is building a place invariant. Therefore the global statement can be formulated that every place invariant of the Petri net theory makes an orientation of the coordinate system necessary [6].

For an integrated specification of automation systems spatial, temporal and causal measures must be unified like in the special theory of relativity where space and time build the four dimensional spacetime. The first step is to find a mapping of the causal axis of processes to another axis. Defining [6]

$$t_k = \frac{t}{\tau}$$

as a causal normalized time specifies the time in seconds for a causal transition on the τ axis. t_k depends on the distance of two states in spacetime. A multiplication of t_k with the causal distance τ leads to the time in seconds which is necessary for the causal transition. By this method, the causal axis of processes can be mapped to another axis. Every system state (eigenstate) can thus be characterized with a 5 toupel $(x_0, x_1, x_2, x_3, x_4)$ where x_0 describes the temporal, x_1, x_2, x_3 the three spatial and x_4 the causal position in a five dimensional generalized spacetime [6].

Velocity of propagation for spatial translations is limited by the speed of light *c*. Because of the embedding of causality in physical spacetime causal propagations are limited by *c* too. The causal axis of processes acts from this point of view like an additional spatial axis [6].

This point leads to the possibility of transferring the geometry of the special theory of relativity to generalized spacetime. The generalized spacetime geometry is determined by the invariance to translations of the quadratic form [5]:

$$q(\vec{x}) = -x_0^2 + x_1^2 + x_2^2 + x_3^2 + x_4^2$$

For specification of the dynamic behaviour of automation systems quantum mechanical methods can be used. The spatial eigenstates of the substances are encoded in the wave function Ψ (not to be confused with the electromagnetic wave $\vec{\Psi}_{EM}$ discussed above), transitions between different eigenstates including selection rules are specified by matrix transition elements $\langle \varphi \mid M \mid \psi \rangle$. $\langle \varphi |$ is a vector of dimension (1, *n*) and is representing the starting state. $| \psi \rangle$ has the dimension (*n*, 1) and represents the target state. The quadratic (*n*, *n*) matrix *M* encodes the selection rules. $\langle \varphi \mid M \mid \psi \rangle$ is equal to the transition probability between state $\langle \varphi |$ and $| \psi \rangle$ [5].

Automation systems consist of discrete causal states (eigenstates). Translations between causal states are determined by selection rules encoded in the Petri net representation of the system. This similarity between quantum mechanical microsystems and automation systems allows the quantum theoretical transferring of matrix transition elements for description of selection rules to causality. Following definition is made. ***M*** **is a quadratic matrix which contains as many rows and columns as the number of system eigenstates. If the causal distance between state** i **and state j is one, then the element** *M (i, j)* = 1 **else** *M (i, j)* = 0. For the causal structure of figure 4 *M* is:

$$M = \begin{pmatrix} 0 & 0 & 1 & 0 & 0 & 0 & 0 & 0 \\ 0 & 0 & 1 & 0 & 0 & 0 & 0 & 0 \\ 0 & 0 & 0 & 0 & 1 & 0 & 0 & 0 \\ 0 & 0 & 0 & 0 & 1 & 0 & 0 & 0 \\ 0 & 0 & 0 & 0 & 0 & 1 & 1 & 0 \\ 0 & 0 & 0 & 0 & 0 & 0 & 0 & 1 \\ 0 & 0 & 0 & 0 & 1 & 0 & 0 & 0 \\ 0 & 0 & 0 & 0 & 0 & 0 & 0 & 0 \end{pmatrix}$$

$\langle \varphi \mid M \mid \psi \rangle = 1$ if the causal distance is one (allowed transition) else $\langle \varphi \mid M \mid \psi \rangle = 0$ [5].

In four dimensional spacetime the continuity equation is [1]:

$$\frac{\Delta Q}{\Delta t} + J = 0$$

The first term specifies the changes of the substance Q per time unit, the second term is the substance current. In generalized spacetime the existence of the invariance of the quadratic form demands an additionally term specifying the causal translations. Using causal matrix transition elements the generalized continuity equation can be written as [6]:

$$\frac{\Delta Q}{\Delta t} + J + \sum_{i=1}^{n}\sum_{j=1}^{n}\frac{\langle \varphi_i \mid M \mid \psi_j \rangle}{\Delta t} = 0$$

In every automation or information system translations in generalized spacetime consider the generalized continuity equation. In many applications the temporal behavior of the system is of great interest, for example in applications with real-time requirements. By evaluating the generalized continuity equation for some subsystems balance equations arise and describe the load of the technical resources. In dependence of this load the scheduling system estimates the distribution of the tasks to the different technical resources [6].

Automation theory is characterized by a given number of operations per volume ΔV and per time interval Δt. This fact implies the existence of a generalized continuity equation including causal translations of material, energy and information resources [6].

The derivation of the generalized continuity equation is won by generalizing spacetime geometry of the special theory of relativity to the world of automation engineering, and by quantizing causal translations by causal matrix transition elements transferred from quantum mechanics [6].

Evaluating the generalized continuity equation for different subsystems leads to the derivation of balance equations. These equations are considered by the scheduling system for the integrated task distribution in the automation system [6].

Thus we have derived a well founded reasoning for definition 2 defining a process in automation systems by using the three functions P, T, and S applied to matter, energy, and information, defining the nine elements of a process [6].

It can be finally concluded that [6]:

Corollary 14: Systems consist of ten physical dimensions according to definition 1, and are governed by the laws of conservation of momentum, angular momentum, energy, and information. Information and causality are Lorentz invariant contrary to momentum, angular momentum, energy, space, and time. Momentum and space, angular momentum and angle, energy and time, and information and causality possess an uncertainty relation, respectively.

Corollary 15: Processes consist of nine physical dimensions according to definition 2, and are governed by three functional elements P, T, and S applied to matter energy, and information. Matter, energy, and information are described in a generalized spacetime and are governed by a generalized continuity equation, respectively.

Corollary 16: Systems and processes define the structure and functionality of automation systems in a physical complete manner by respecting all relevant physical parameters: momentum, angular momentum, energy, information, space, time, and causality. Hence, said systems and processes define automation theory in a complete manner.

Bibliography

[1]: Carlo Maria Becchi; Massimo D'Elia: Introduction to the Basic Concepts of Modern Physics; Special Relativity, Quantum and Statistical Physics; Third Edition; Springer; 2016.

[2]: https://en.wikipedia.org/wiki/Planck_units#Cosmology

[3]: C. A. Petri: Nets, time and space; Theoretical Computer Science 153; 1996.

[4]: A. Mircescu: Physical Definition of Information and Causality, their Special Relativistic and Quantum Mechanical Structures, and the Law of Conservation of Information; GRIN: Catalog Number V343915, 2016: ISBN: 9783668365209.

[5]: A. Mircescu: Über die Beschreibung und Optimierung verteilter Automatisierungssysteme, Doctoral Thesis, Technische Universität Carolo-Wilhelmina zu Braunschweig, 1997.

[6]: A. Mircescu: Automation Theory Defined by Systems and Processes; GRIN: Catalog Number V351220, 2017: ISBN: 9783668378780.

[7]: Lüke: Signalübertragung, 5. Auflage, Springer-Verlag, 1992.

[8]: Y. Aharonov and D. Bohm: Time in Quantum Theory and the Uncertainty Relation of Time and Energy; Physical Review, Volume 122, Number 5, June 1 1961.